NATURE'S BEAUTY IN COLORS

Rediscover the Joy of Coloring with Breathtaking Landscapes

This book belong to

_____

_____

Preface

Welcome to "Nature's Beauty in colors: Rediscover the Joy of Coloring with Breathtaking Landscapes"! In this coloring book, you'll find 70 stunning natural landscapes, each waiting for you to bring them to life with color. From misty forest trails to snowy peaks, rolling vineyards to ocean sunsets, these scenes are sure to inspire your creativity and awaken your sense of wonder.

As adults, we often get caught up in the daily grind and forget to take a moment to appreciate the beauty around us. Coloring is a wonderful way to slow down, quiet the mind, and reconnect with our inner selves. It's also a fantastic way to explore our creativity and play with color in a way that feels both relaxing and fulfilling.

In this book, you'll find a wide range of landscapes, each with its own unique mood and energy. Some scenes are bold and vibrant, while others are more peaceful and contemplative. But all of them share a common thread: the beauty and majesty of the natural world.

We hope that coloring these scenes will bring you a sense of peace, joy, and connection to nature. So pick up your favorite coloring tools, find a comfortable spot, and let yourself get lost in the beauty of these landscapes. Happy coloring!

Nature's Landscape

A natural landscape is a geographical area that has not been significantly altered by human activity. It is characterized by its raw and unspoiled beauty, with features such as mountains, forests, oceans, and rivers that have been shaped by natural processes over time. Natural landscapes are often admired for their aesthetic qualities and can evoke a sense of wonder, awe, and tranquility in those who observe them.

One of the most striking features of natural landscapes is their diversity. From the towering peaks of the Himalayas to the serene beaches of the Caribbean, natural landscapes can take on a wide range of forms and provide a variety of experiences to those who explore them. In addition to their visual appeal, natural landscapes can also provide opportunities for recreation, education, and scientific research.

Natural landscapes are often appreciated for their ability to evoke emotional responses in humans. The grandeur of a mountain range, the tranquility of a forest, and the power of a thundering waterfall can all evoke a sense of awe and wonder in those who witness them. Many people find that spending time in natural landscapes can be a source of inspiration, relaxation, and renewal.

The beauty of natural landscapes has long been celebrated in art, literature, and philosophy. Throughout history, artists have attempted to capture the essence of natural landscapes in their work, while writers and poets have found inspiration in the beauty of the natural world. Philosophers have also reflected on the relationship between humans and the natural world, and the role that natural landscapes play in shaping our understanding of ourselves and the world around us.

Despite their many benefits, natural landscapes are also threatened by human activity. Climate change, pollution, deforestation, and overdevelopment are just a few of the ways that humans have impacted natural landscapes around the world. Efforts to protect and preserve natural landscapes have become increasingly important in recent years, as people recognize the value of these areas for their beauty, ecological significance, and cultural heritage.

In conclusion, natural landscapes are a source of beauty, inspiration, and awe for many people around the world. Whether they are admired from afar or explored up close, these areas remind us of the power and majesty of the natural world and the importance of preserving it for future generations.

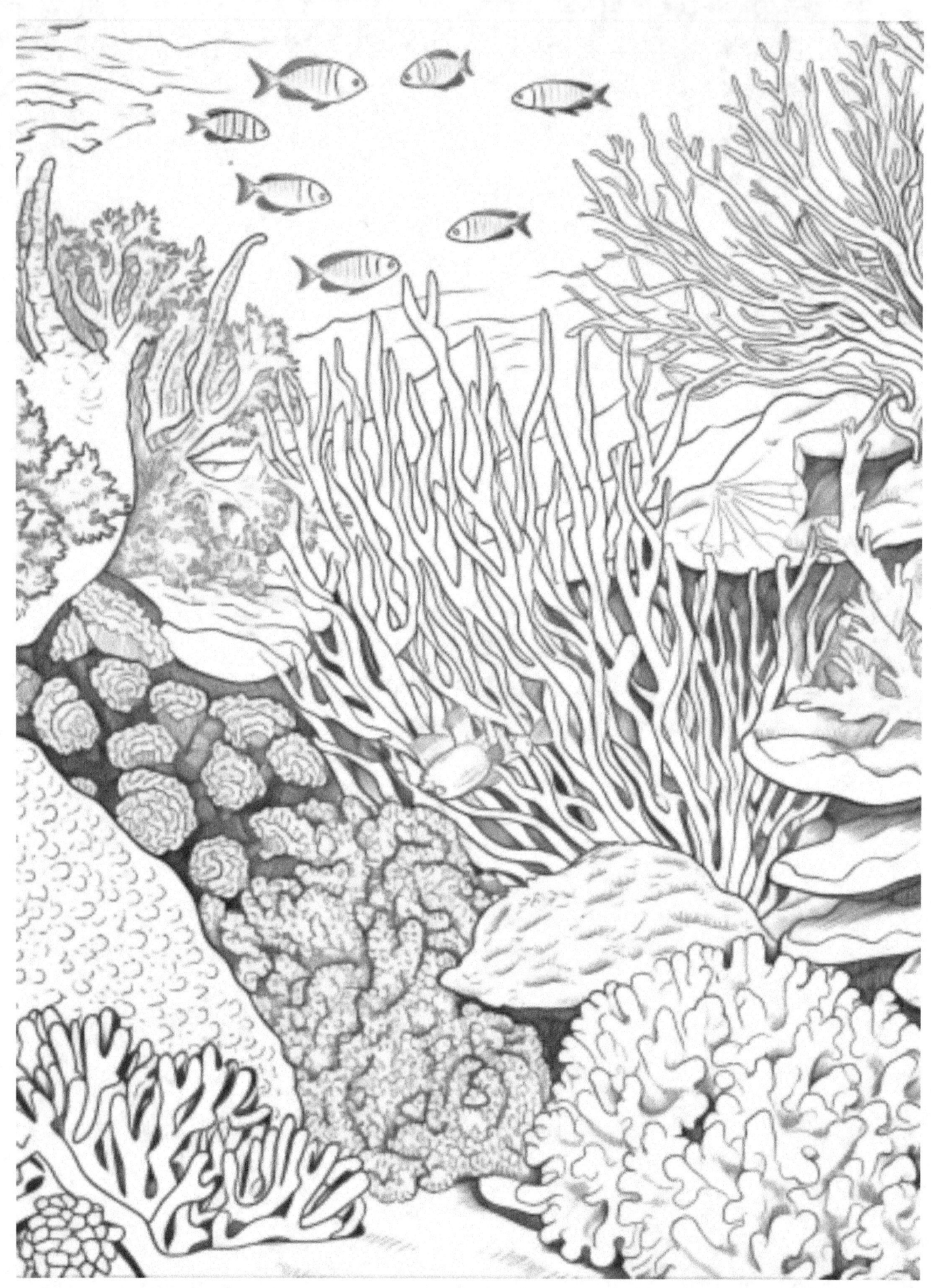

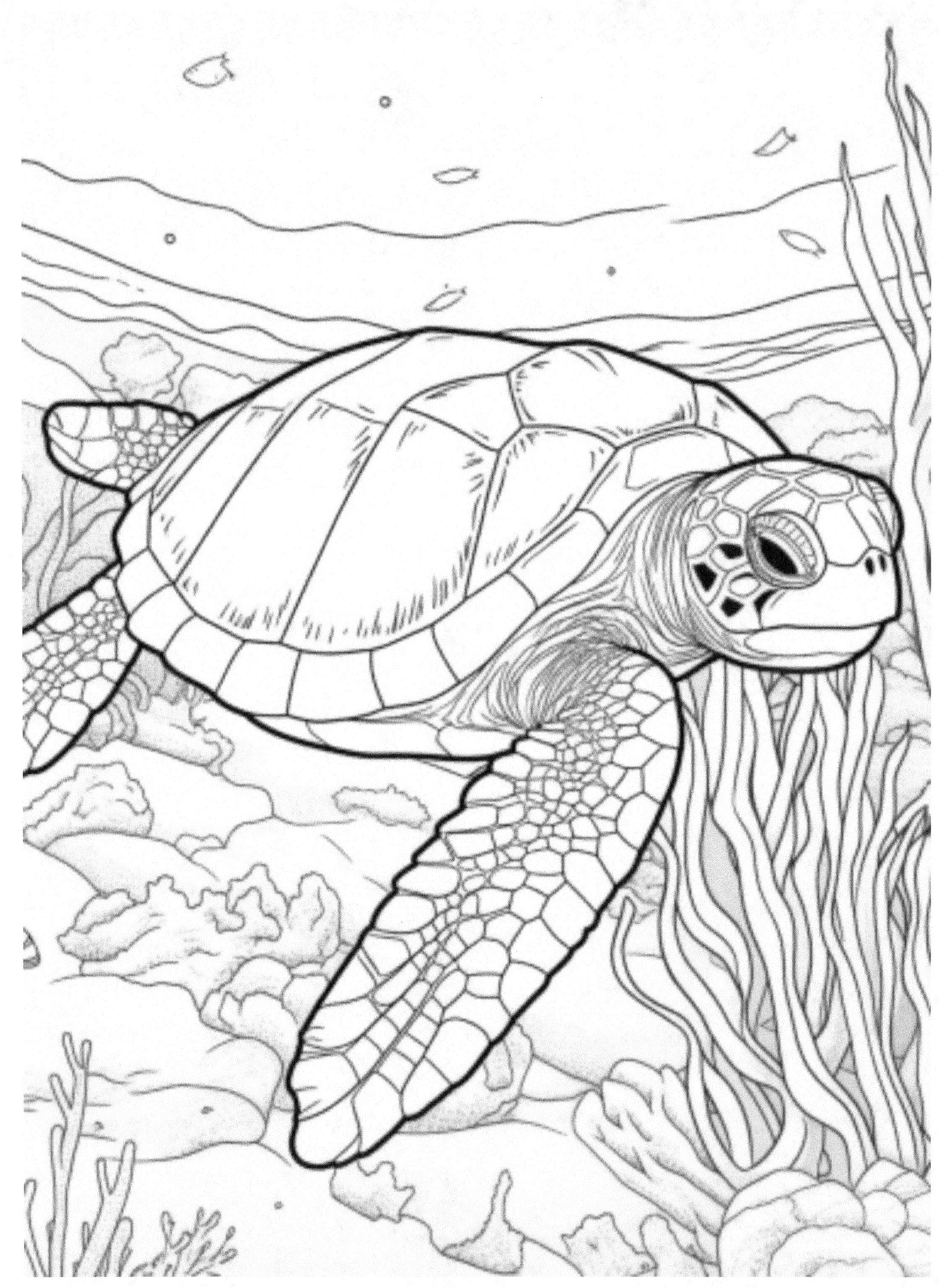